ODDS N ENDS IN RHYME

THE LIFE AND WRITINGS OF
WALTER PHILLIP "PHIL" JORDAN JR.

ODDS N ENDS IN RHYME:
The Life and Writings of Walter Phillip "Phil" Jordan Jr.

Copyright © 2021 by Walter Phillip Jordan Jr.

All rights reserved. No part of this book may be reproduced or utilized in any form or by any means, electronic or mechanical, including photocopying and recording, or by any information storage and retrieval system, without permission in writing from the publisher.

First Edition

The views expressed in this work are solely those of the author and do not necessarily reflect the views of the publisher, and the publisher hereby disclaims any responsibility for them.

Published by Tactical 16 Publishing
Monument, CO

ISBN: 978-1-943226-62-7 (paperback)

PUBLISHER'S NOTE

All of the poems in this book were composed during a lifetime of military and law enforcement service. Each piece serves as a true expression of the experiences of Walter Phillip Jordan Jr., better known to family and friends as "Phil." The poems touch on both his struggles and triumphs, ranging between humorous to dark subjects while always feeling personal and relatable.

None of the poems in this book are typeset using a "justification" setting. When writing the poems, Phil counted out the total number of characters for each line (including all of the letters and spaces). He carefully crafted each piece so that the lines are completely even. As Phil was creating each poem, each stanza had to rhyme, make sense within the poem, *and* it had to match the line length count for that piece. It is a true masterpiece, and his family hopes that you will recognize this book for the amazing work of art that it is.

Phil wrote these poems and wanted his book published to offer hope to the thousands of veterans that struggle with post-traumatic stress disorder (PTSD), which he also suffered from for many years. In one of the poems, he tells his family that his PTSD developed during his military and law enforcement surface; however, it wasn't until the 1980s that it fully surfaced. According to Phil, writing poems and fishing helped him to "combat the ghosts of the past." His family hopes that reading his works will help others as well.

WANTED
(NEW SHERIFF IN "RHYME" ARIZONA)

Endowing of author to fulfill a need,

Poetic emotion on parchment to **bleed**.

Scribe **inked** art, alphabet to assail,

Flay of pen **point** opt word to impale.

Periodically crux is fast river flow,

Is normally logjam & is glacier slow.

Words universal and **thoughts** his own,

Harvesting **dreams** as yet to be known.

Written picture artistically **painted**,

Reading or reciting less complicated.

Pen running dry your brain to engage,

Opt, **close** the book or **turn** the page.

BLACK OR **COLORISTIC** POEM FORMAT: PERSEVERE
EQUAL SUMLINES. (COUNT 'EM)

Dedicate to my favorite daughter Jane Evelyn Breshears,
Persistent in nagging me and nagging me for many years,
Just wrote down something that told the story about me;
Love for you and your three brothers here is   .

GENEALOGY

Pre-1984, Walter "Phillip" Jordan, Jr., was legal name,
Notice my birth certificate didn't quite read the same.
Now in the following history that I am about to relate,
You can re-live my life with me & anticipating my fate.

The above name was logged when I was born and baptized;
School, military, jobs, married & any papers legalized;
Born 3-22-27 Phoenix, Arizona, far out into the sticks,
1613 East Brill outside the city limits, country hicks.

With parents and my half brother, why still feel alone?
Four different houses & was never comprehended at home,
Weekending with many Aunts and Uncles one or the other;
Spending summers in California with my father's mother.

First team in the major sports & above average student,
A good sense of humor, popular, reasonable and prudent,
Bashful around girls, poet and a pet of multi teachers;
Lover of life, of nature & all of God's many creatures.

15 year old San Francisco shipyard worker World War II.
At 16.*cranked planes for Chinese Army cadets who flew;
17TH birthday & 20 days enlist WWII Navy blue 4-11-44,
Age 18 *5 South Pacific battle stars, 2-21-46 exit war.

Age 19 re-enlist 7-30-47 can't adjust to civilian life,
Serving three years as Army paratrooper; still no wife;
Trained as construction ironworker for almost one year,
7-6-51 entering law enforcement which I made my career.

I fell in love and took for my wife Portia Condie Call,
She is dear mother for you and your three brothers all,
22 years of varied law enforcement is a very long time,
Will condense and share some experiences that are mine.

County Attorney Investigator, Adult-Juvenile probation,
*B.I.A. Federal Investigator on San Carlos Reservation,
Deputy, Sergeant, Lieutenant, Chief Deputy and Sheriff;
During this time I paid full price, plus health tariff.

*Thunderbird Airfield, Glendale AZ
*Verification of 3 more Bronze Battle Stars in progress
*Bureau of Indian Affairs & San Carlos Apache Indian "Chief"
of Police by Tribal appointment.

Qualified tear gas, motorcycle, horse, four wheel drive,
Car, shotgun, pistol, machine gun, few tools to survive;
National Sheriff Institute and *F.B.I. National Academy,
Variety of law enforcement schools & Jr. College degree.

A Personal request of condemned prisoner #19795 in 1959,
Arizona State Prison witness him executed for his crime;
1966 Kingman, Arizona in shootout my opponent laid dead,
1969 the White House* President Nixon and I broke bread.

Commendation letters, plaques, trophies memory of award,
Appointed, elected fraternal offices too many to record;
Total experiences whether they were either happy or sad,
Reflected on how I raised you or guided you as your dad.

Do you remember that in 1974 while in a hospital I died?
When our Lord sent me back you were all near my bedside;
I returned a different person and a little bit confused,
There is a job I haven't done or talent I have not used.

In 1984 while in the process of seeking my own identity,
Many of future goals changed between your mother and me.
32 years to the day, my faithful wife, lover and friend,
Divorce has put our productive marriage to peaceful end.

My marriage and birth of you and each of the three boys,
Were the highlight of my life and gave me the most joys.
*P.T.S.D. military & law enforcement burden me as a man,
Camouflage trauma; try to be the best father that I can.

To each of the four of you with whom I share my surname,
It is proudly given to you and unblemished of any shame.
My 3 sons, some times you did things that hurt my pride,
But neither my name nor my love none of you were denied.

I do not know what time is left, or what there is to do,
But I desire making it clear to each of the four of you,
I could stand tall on top of the world and loudly shout,
"LOVE YOU INDIVIDUALLY"! Regardless of how you turn out.

Presently am 240 lb., 6'2", blonde, blue-eyed, young 57,
Knowing that when I die again that I am going to Heaven;
Name's correct except "Philip" is single not double "L",
You know I am going to Heaven because there is no "Hel".

*F.B.I Federal Bureau of Investigation
*P.T.S.D. Post Traumatic Stress Disorder

THERAPY
(*P.T.S.D.)

2 military hitches include WWII Vet,
Violence continues to follow me yet.
22 years, in & out of *prison/jails.
In prime of life as my health fails.

Greet each day with more difficulty,
I accept social security disability.
Despondent & more restless each day,
Work since 15 & never learn to play.

Doctor, I quest medicine to help me,
No Pills? Choice nut house or hobby?
Found hobbies a disabled man can do,
Qualify in one & other is brand new.

My new hobby that I choose to learn,
Fish artificial lure, drown no worm.
Tournament or fun I catch & release
Mother Nature & I embraced in peace.

Play cards & wearing my fishing hat,
Tell my story to players as we chat.
A player for whom I have little use,
Is ranting & raving in verbal abuse.

Gist, mental therapy success for me,
Does not justify my fishing cruelty.
Eyeball critic & say in stern voice,
DRIVE-BY SHOOTING WAS MY 2ND CHOICE!

TELL ME QUIT FISHING FOR MY THERAPY!
CAN ACTIVATE 2nd CHOICE IMMEDIATELY!
HEAD YOUR WAY AS I START TO DEPRESS!
PAY YOU A VISIT I KNOW YOUR ADDRESS!

Therapy debate cease need to defend,
Critic? Kisses my butt to be friend.
Game over, we all exit to our homes,
In reality, 2nd choice? Write poems.

*Post Traumatic Stress Disorder,
*Arizona Law Enforcement
*WWII we had battle fatigue,
later changed to P.T.S.D. I did
not get any VA help till 1998.

TEENAGER
(WORLD WAR II)

The following facts that are near to unfold,
Great test for memory of a man 91 years old.
Time's 76 years past, names & faces are dim,
People who forfeit to make world safe again.

Washington D.C. three weeks till Xmas, 1941,
USA hosts peace envoy of Land of Rising sun.
Emperor Hirohito Cipango God? Or mortal man?
Pacific reveals atrocious Satan rules Japan.

Blatantly void God & mortals peace on earth,
Violated Sabbath in month of Christ's birth.
12-7-41, tranquil Hawaii Sunday & post dawn,
Murderous aircraft attack by Sons of Nippon.

Strafe airfields, barracks and Pearl Harbor,
Messengers of Death emit declaration of war.
Mikado! 14-year-old boy vows, not yet a man,
I'll witness victory and surrender of Japan.

Arizona runaway & high school freshman year,
San Francisco pinsetter while seek a career.
Shipyard coppersmith, division of submarine,
Law says, school one day a week till age 16!

Rich man called Uncle Sam but not blood-kin,
Inviting me to take extensive tour with him.
Offers salary, job training & paid vacation,
Room, board, retirement, cash for education.

Those who do not comprehend Navy vocabulary,
To assist you I now provide a mini glossary.
Swab-wet mop, starboard-right, ladder-stair,
Shore Patrol-Navy Police skivvies-underwear.

Deep 6-sea bottom, fish-torpedo, deck-floor,
Line-rope, forward or bow-front, hatch-door.
PT-patrol torpedo boat, O-D-officer on Deck,
Boot-camp or recruit-short arm-pecker check.

Watch-on duty, fantail-wide deck, left-port,
Topside-above a deck, D.E.-Destroyer Escort,
ASW-Anti Submarine Warfare engage if detect,
Mate-companion, convoy or escort-to protect.

Turn to-duty time, chow-food, bulkhead-wall,
Kaiten-human torpedo, dining area-mess hall,
Depth charge-bomb neath sea via K gun, rack,
Picket-patrol alert prevent surprise attack.

Ship-boat, R&R-rest & recreation, rank-rate,
Hedgehog-62 pound mortar shell 24 aggregate,
Muster-assemble, Joe-coffee, liberty-ashore,
Overhead-ceiling, Pearl-deep 6 **grave** harbor.

Detect by image-sea surface, airborne-radar,
Underwater detection by sound signals-sonar,
Dog fight-aircraft combat U.S. Vs Jap enemy,
Jap suicide attack, boat plane-sub-Kamikaze.

Flattop-aircraft carrier, V-J Day-end WW-II,
20mm-40mm & 5"-gun caliber, propeller-screw,
Aft or stern or fantail-rear, pilot-Airdale,
*****Zoot suiter**-flamboyant clothed Mexican male.

Underage, Dad sings my enlistment into Navy,
San Diego boot, then Los Angeles USN Armory.
Sunset & Figaroa, Radio School 19 week stay,
Stateside war zone & without any combat pay.

SO-caliber machine gun backs our front door,
Guard duty, Armory walls bullet holes score,
Navy and **"Zoot Suiters"** combat hand to hand.
L.A. liberty as Shore Patrol, wear arm band,

Learn to liberty famous "Hollywood Canteen."
Movie stars entertain. A serviceman's dream.
Finish Radio School and graduation complete,
BEWARE "TOJO" I'm en route to Pacific Fleet.

New home, USS KENDALL CARL CAMPBELL, DE-443,
14 officers, 172 crew, 1 juvenile sailor-me.
Multi-medaled Airdale & killed at Coral Sea,
Ships valor-namesake to carry us to victory.

Wasn't a shipmate during East Coast history,
Prior to San Diego, to me remains a mystery.
Saga of Captain's sacred! ice cream machine,
Machine can't swim, deep-6 never again seen.

*The Zoot suit was a long coat with padded shoulders,
full pleated trousers cuffed tightly at the ankles,
and a heavy keychain adangle.

Soldiers, sailors, and marines despised zoot suiters.
Just the wearing of said costume invited a brawl,
always dangerous, sometimes deadly.

My ship's in San Diego awaiting at the dock,
Sea bag on my shoulder, up gangplank I walk.
Smokestack insignia "Goofy" clad Navy white.
Astride torpedo hold depth-charge for fight.

Century yards length and 3½ dozen feet wide,
Twin screws, draft XIV feet, DE emits pride.
Ship arms 5 inch, hedgehog, dual 40 forward,
3 fish tubes midship-20s 5 port 5 starboard.

Dual 40, 5 inch, 8 K guns, 2 D.C. racks aft,
Navy's SMALLEST, ALL-PURPOSE FIGHTING CRAFT.
Salute flag, O.D., I request to come aboard,
Thoughts? Please! Bless this ship Dear Lord.

Over torpedo locker & separated by bulkhead,
5-inch ammunition storage aft & my bunk bed.
Black shipmates do their job the same as us,
Why? Quarters in rear of our watercraft bus.

Rate 3/C Petty Officer on Uncle Sam's Yacht,
$71 add 10% for overseas, monthly pay I got.
Altho tragedies of WW-II continue to remain,
Share some highlights won't reveal any pain.

Adios Hawaii, pass horror of **Battleship Row**.
Image etched in my memory, off to war we go.
V.M. Conkle asleep? Movie theater in Hawaii?
Missed ship! Balance of war listed absentee.

Route combat zone via Hawaii to Philippines,
We escort, picket, convoy & anti-submarines.
U.S. Troops fighting in air, on land & seas,
Repel enemy by exchange of multi **casualties.**

Seek submarines, midget, kaiten & full size,
Aircraft attack out of the sun, set or rise.
Baka pilot, planes-Betty-Zero-Judy-kamikaze,
Visit us enough quickly learn each identity.

NOW-HEAR-THIS! NOW-HEAR-THIS! ALL-HARDS-HOW!
EITHER-turn to, smoking lamp, standby, chow.
*Tokyo Rose, all clear, fuel & ships stores,
Payday, mail, lights out, all sports scores.

Sweepers! Man your brooms-short arm-liberty,
Calisthenics, worship, blackout, work party.
All hands on deck, R&R, enter or leave port.
WELL DONE, **fire** or GQ drill, weather report.

Dump trash or garbage, *bos'n pipe, laundry,
Secure, drop or aweigh anchor, repeat movie.
Muster, sick call, inspection, updating war,
Loudspeaker **BLASTING** plus a whole lot more.

Chow is good, sometimes meat a little green,
Floured weevils in chow new form of protein?
GQ-General Quarters call to battle stations,
Hot-Joe & baloney sandwiches our GQ rations.

Ship, either side of International Dateline,
Safely destroy 36 of any type floating mine.
Typhoon wrath, turbulent sea, ship 55°list,
Web bunk in line, watch & chow perils exist.

P.T. boats join us for bread, movies & dine,
Did we host John F. Kennedy and crew PT-109?
Underwater sonar contact submarine or whale?
New contact dolphin or Jap-fish? We prevail.

Hot beer or cola, R&R desolate Mog-Mog Isle,
Juices *torpedo & grapefruit R&R worthwhile.
"DE's", war correspondent Ernie Pyle to say,
Crew entitled **BOTH** submarine **AND** flight pay.

*Tokyo Rose - Japanese lady radio broadcaster
*Bos'n, Job Title - Pipe, Flute-type whistle call
*190 proof alcohol used as torpedo propellant

Shock! President Franklin D. Roosevelt died,
His son FDR Jr. captains DE-442 at our side.
Shark watch, ships crew swimming in the sea,
Few times rifleman job is my responsibility.

Dud Jap-fish? Shaft screw, Ulithi in repair,
U.S. & Japanese planes dog-fight in the air.
From seas rescue 38 downed Airdales or crew,
Flattops reward us ice cream as a thank-you.

Shore battery target classify us expendable,
Continue duties, name us LUCKY & dependable.
Iwo Jima, witness a moment of world history,
U.S. Troops, American Flag, Mount Surabachi.

Kerama Retto "Ghost", radar says solid mass,
Back & Forth thru "Ghost" in safety we pass.
Okinawa and multi other islands in vicinity.
69 days at sea, my Dad died & unknown to me.

Message missed by a radioman, name not mine,
Crew emancipated, Captain Johnson, reassign!
Bagley passed fantail as boating him ashore,
Crew threw garbage & 2nd pass we threw more.

Awe at great white shark circling PBY plane,
Rescue crew, & sink hazard in shipping lane.
Next to last sea battle 5 days to V-J WW-II,
ASW resolve sink 3 Jap subs duties continue.

Raids on Empire Japan July 1945 we complete,
ASW Tokyo Bay as Japan surrenders in defeat.
War duties more than some, less than others,
Allied victory **WELL DONE** sisters & brothers.

Against E ire of Japan ORDER cease offense,
Continue ASW to engage only in self-defense.
4 on and 4 off watch now is no longer valid,
Sleeping 8 hours dream Mexican food & salad.

Homesick & awaiting our time to return home,
Islands & Seas of Pacific, continue to roam.
ASW duties for unarmed vessels of variation,
Carrying troops or equipment for occupation.

Subic Bay Dry dock & maintenance is overdue,
Any port seagull crap polka dot ship & crew.
Resume ASW, occupation now 80 days of peace,
ORDER! Via Yokosuka, Pearl, USA, duty cease.

USS Arizona in MASSACRE unit **Battleship Row**,
Future memorial, Arizonan, tears still flow.
Rip Van-Conkle awakened from Hawaiian Dream,
Aboard as baggage not crew in fighting team.

Ship will be Stateside by Thanksgiving 1945,
Blessings for the **dead** & thanks to be alive.
8th grade education, now 18 & legally a man?
Unprepared for adulthood, adjust best I can.

EXIGENT WWII MEMORIAL
(Written Prior to WWII Memorial)

In Washington D.C.,three weeks till Christmas 1941,
USA hosts two peace envoys from land of rising sun.
Expose emperor Hirohito Cupango god? Or mortal man?
Pacific reveals atrocious son of Satan rules Japan.

12-7-41 tranquil Hawaii, sleepy Sunday & post dawn,
Murderous aircraft attack by sneaky sons of Nippon.
Bomb & strafe airfields, barracks and Pearl Harbor,
Coward messengers of death emit declaration of war!

Amid billow smoke battleship-bridge waterlines sea,
America image & battle standard in Pacific history.
Postwar, Pearl Harbor Hawaii is American Fiduciary,
Impregnable USS Arizona 900 plus fatality **mortuary**.

Violate USS Arizona sanctum, **tomb of eternal sleep**,
By water-taxi churning noise, sea, oil & sand deep.
Trespass footsteps and voice echo's via deck awash,
MURDERED Sailors/Marines cry, **'ENCROACHMENT QUASH'**.

Enough! 'ABANDON SHIP' & move etched enmity ashore,
Honor souvenir with quiescent retirement ever more.
Asiatic or European theater of military operations,
Military blend creed, gender, color, race, nations.

Defending with valor, distinction & American pride,
16,112,566 Serving, 670,846 *Wounded, **405,399 Died.**
Cornerstone WWII memorial resurrect severed-bridge,
Sans WWII veterans honor memorial, CIVIL SACRILEGE!

*Non fatal

VIRGIN

Midst barn odors and fresh cut hay,
We stand committed, nothing to say.
Alone in the barn, just you and me,
Stroke her bare body repetitiously.

Patted on her butt to her shoulder,
Makes no difference that I'm older.
This's something we both had to do,
First time for me, not new for you.

Her eyes are big, beautiful, brown,
She nuzzles my neck as I bend down.
I feel her warm breath on my cheek,
My knees are getting terribly weak.

On my brow beading, nervous, sweat,
My roving hands eager, clumsy, wet.
Quiver as I fondle her tenaciously,
She is tolerating me so graciously.

Near me is voice, "Doing okay Son?"
Reply, "About to get the job done!"
Grabbed her teats, pulled then WOW!
Today's the day I first milk a cow.

SORROW

Passenger on Greydog mournfully reminiscing,
A pickup truck, trailer & horse I'm missing.
My orphaned horse that I raised from a colt,
Thought of him **dead** triggers emotional jolt.

Pulled into hometown bus station, near **dark**,
Grab my gear, saddle over shoulder & embark.
First aid station around corner and not far,
Rodeo, horse racing is working cowboys' bar.

Like an emergency I come fast thru the door,
Belly up to bar my tack & gear on the floor.
Throw a fifty on the bar & call SHOT & BEER!
Beertender have a sad story for you to hear.

Now bankrupt, devastated & quivering inside,
Raise from colt my horse & future just **died**.
Tend & baby this horse in sickness & health,
Natural runner trained for winning & wealth.

Quarter horse & faster than lightning flash,
Cinch to win, intend to bet big wad of cash.
Trailer horse to current Arizona State Fair,
Know horse racing & betting now legal there.

Horse glides in 2nd place runs next to rail,
Nose locks under in-season lead mares' tail.
My horse soon forgetting he has been gelded,
Finishing second as if to her he was welded.

Sell my pickup & trailer then honor my bets,
Image of special horse cowboy never forgets.
50 bucks in shots & beer: horse I toast you,
Packinghouse money horse is dog chow & **glue**.

EXODUS

Sitting on the toilet doing my due,
Siamese kitten approaching mew-mew.
She purrs and then arches her back,
Disappeared at sound of gas attack.

Noise is offensive and so is smell,
Brown vapor rises, floats and fell.
Without disclaimer would be remiss,
Ate nothing that smelled like this.

Nostrils burning, eyes tear galore,
Check new exit hole in screen door.
Gas propelled kitten cleared fence,
Vapor trail! Kitten not seen since.

SALES
(Thanks Dale Carnegie)

Must provide security and necessities of life,
Not only for myself but for children and wife.
What job provides potential and pay the price?
Cost of living, retirement, what will suffice?

First I would select a challenging occupation,
That's in demand throughout the entire nation.
What's this occupation that I'm searching for?
One that enables me to reach my goals or more.

Research many want ads, advertisements, books,
Neglected present job and receive dirty looks.
Worried, lost sleep, until I began to realize,
SALES! That is what will bring home the prize.

There isn't any reason for me to tell any lie,
A measure of success will be easy to identify.
If I enthusiastically follow a few good rules,
That provide salesmen with real selling tools.

Stop!------------------------do nothing more!
If unable to pass through your own front door.
Select a quality product which I want to sell;
Learn this product, believe, remember it well.

Desire, enthusiasm and organizing all my work,
Leaves me a routine that's difficult to shirk.
Know about my prospect, his product, his need;
With this knowledge, I am prepared to proceed.

First, get his attention, arouse his interest,
Then put my facts and many benefits to a test.
Done with sincere enthusiasm, before he knows,
He not I has very successfully made the close.

Thanks I know we each made a brand new friend,
Both made a beginning of which there's no end.
But back to work I can not afford to hesitate,
When retirement arrives don't want to be late.

VIGILANCE PARK

Having a very relaxing day here in this beautiful park,
Guess I better leave now for it's rapidly getting **dark**.
Let me see now, keys, towel, thermos, **sunglasses**, book,
Suntan lotion, blankets, picnic basket, one final look.

Quantity of items to carry thank goodness it's not far,
Reminiscence of this day walking reluctantly to my car.
Under a **shady** tree I spread my blankets upon the **grass**,
Read a book, ate lunch and drank iced tea from a glass.

Napped, went strolling, fed the pigeons, picked a rose,
Waded in the pond enjoyed the mud ooze between my toes.
Threw bread to noisy ducks and frog jumped from fright,
Observe boy climbing tree to retrieve his trapped kite.

Slippery sidewalk, result of leaking drinking fountain,
Group of boys on the hill playing king of the mountain.
Motherly little girls playing with their various dolls,
All kinds of **graffiti** on the public toilet **dingy** walls.

Dogs running off their leash, music from the park band,
Gum on the park bench and couples walking hand in hand.
Volleyball, swings, softball, a hot air balloon flying,
Football, Frisbee, slide, teeter-totter, babies crying.

Parallel bar, hopscotch and white clouds in a blue sky,
Jumping rope, playing in the sandbox, birds on the fly.
Here is the car; I'll put picnic supplies in the trunk,
One of these days, need to clean out some of this junk.

The guy in a overcoat is acting strange and I bet cash,
Yep, opened his coat, he's nude, that's called a flash.
Mr. Flasher stop-stop, come back please don't run away,
Wait let me introduce myself I'm your neighborhood gay.

(Poetic AWARE intent, neither endorsement, slur nor lifestyle.)

CITY

Led my hobbyhorse outside,
Want to take a happy ride.
Profiled the painted head,
Highlighting face, on red.

Horse body, painted board,
Reins of heavy nylon cord.
Mane of leathers shredded,
Wheel for tail & embedded.

Super Z-Ray automatic gun,
Armed & now ready for fun.
Ride with wind in my face,
Try saving the human race.

Eyes shaded by cowboy hat,
On my horse I fidgety sat.
Home! Fastest I ever rode!
Horseless, pee, a commode.

COUNTY

Stick horse waits outside,
As owner burst with pride.
On broom, coffee can head,
Hills brothers, color red.

Castoff broom, horse body,
Paint peels, dull, shoddy.
Neckerchief replaces mane,
Double shoelaces for rein.

Holstering trusty six-gun,
Armed & now ready for fun.
Ride with wind in my face,
Try saving the human race.

Eyes shaded, baseball hat,
On my horse I fidgety sat.
Bladder! About to explode!
Horseback, pee, dirt road.

CITY/COUNTY

Led my hobbyhorse outside,
Want to take a happy ride.
Stick horse waits outside,
As owner burst with pride.

Profiled the painted head,
Highlighting face, on red.
On broom, coffee can head,
Hills brothers, color red.

Horse body, painted board,
Reins of heavy nylon cord.
Castoff broom, horse body,
Paint peels, dull, shoddy.

Mane of leathers shredded,
Wheel for tail & embedded.
Neckerchief replaces mane,
Double shoelaces for rein.

Super Z-Ray automatic gun,
Armed & now ready for fun.
Holstering trusty six-gun,
Armed & now ready for fun.

Ride with wind in my face,
Try saving the human race.
Ride with wind in my face,
Try saving the human race.

Eyes shaded by cowboy hat,
On my horse I fidgety sat.
Eyes shaded, baseball hat,
On my horse I fidgety sat.

Home! Fastest I ever rode!
Horseless, pee, a commode.
Bladder! About to explode!
Horseback, pee, dirt road.

LADY

Labor of love to multi negate chore,
Obsessive compulsion welcoming more.
Vaccinating all with love injection,
Encompassing tender care, affection.

Solicitously opens resources to all,
Witching precepts pain to forestall.
Embrocating for pleasure for health,
Enjoyment of giving is great wealth.
Treasure to be accepted for a blend,

Acknowledging to you my best friend.
Notorious and known without compare,
Natural heartbeat of perpetual care.

DECEMBER LETTER

1984, divorce, 32 year marriage unchained,
Unknown: Enough time left to be retrained?
Nearly drown, flunk test to walk on water,
12 grandchildren, 3 sons and one daughter.

School of hard knocks & court case degree,
Imprisoned by employment now totally free;
Arizonan has Russian fingers, Roman hands,
Massage skills dormant & awaiting demands.

Toes 2 toes, nose 2 nose, we are 6 foot 2;
Beautiful eyes colorable bloodshot & blue;
Crewcut blonde hair copies military style,
Impish grin interlaces with dimpled smile.

Sumo physique, wrestling talent is denied,
Waiting film accept my image, camera died;
No competition to any handsome movie star;
Drive an old truck, don't fit a fancy car.

Country hick sterile in a very small town.
Board games, cards, fishing, admit renown,
Any snoring contest winner of first prize.
Newspaper obituary void of my name, arise.

Plausible rogue, march & beat my own drum.
Replace city clothes & garb as cowboy bum.
Shirts are wrinkled and holes in my socks.
My favorite drink is Geritol on the rocks.

My get up and go has done got up and went.
Not broke but bank account seriously bent.
Dowry? Gold teeth, lead ass & silver hair.
Quit tobacco, with drugs never had affair.

Gourmet chef, create a masterpiece to eat:
Pork & beans, baloney, hot shredded wheat.
Sometimes knees buckle, my belt is unable.
Have charleyhorses but don't own a stable.

Telephone silenced. Ears continue to ring.
Bullfrogs croak, donkeys bray when I sing.
My teeth in water smile through the glass.
Can't dance, can intermission first class.

Mind making appointments body cannot keep,
Friendship to sow, pray companion to reap.
Provide this information to you Mr. Claus,
Merry Christmas me with **6** PAC of Grandmas.

PURPOSE

Silhouette to the December sky,
Parasite in tree branches high.
Evergreen mistletoe did dangle,
The life of tree will strangle.

Yellow flowers & white berries,
Harvest as custom of centuries.
Green leaves, stem, tied a bow,
Christmas tradition as we know.

Over person's head & suspended,
Mistletoe, motivation intended.
Enjoy! Embrace people kissably,
Nature's reward to save a tree.

GRAMPY

18th century, April twenty eight, eighteen ninety six,
Day is beautiful in Afton, Wyoming, out in the sticks;
A healthy baby boy born to be named Cyril Alfred Call,
Had many names but was best known as Grampy to us all.

His childhood was pioneer, so poor and full of strife,
He obtained his education before he took a young wife;
After raising responsible children that numbered four,
Parents had different opinions and parted forevermore.

Life went on apprehensively, lonely and after a while,
He met another pretty lady that generated him a smile;
They are married after proper courtship and instantly,
Each contribute toward a very large ready-made family.

On April twenty-eighth, nineteen eighty-four we cried,
For on his eighty-eighth birthday Grampy quietly died;
Funeral service's attended by family and many friends,
This period is where our fond memory of Grampy begins.

Husband, father, brother, a traveler of far off lands,
Diamond ring, father-in-law, teacher, walked on hands;
Motorcycle, hypnosis, nuts, uncle, mission, Cadillacs,
Black jellybeans, movie, grandfather, adjusting backs.

High Priest, enchilada, skater, vitamins, color's red.
Doctor, stepfather, sawdust, requires a king-size bed?
Mustache, energy, bad driver, whistler, maker of toys,
Great grandfather, scholar, situps, hearing aid noise.

Noon nap, photographer, acupuncturing, telephone talk,
Taurus, workshop hats, friend, short legs, brisk walk;
Pool player, fresh fruit, Mormon, philosopher, cheese;
Know he was a lot more but was certainly all of these.

Grampy preceding all of us to skillfully pave the way,
To aid with our transition on that final judgment day;
Deliberately left behind his fortune & great pleasure,
As he personally gave each of us memories to treasure.

WINNER
(Narrator-Lucifer-Cowboy)

Creature person traveled up from afar,
Silhouetting in doorway of cowboy bar.
Buttoned B.V.D. union suit? Color red?
One horn top front each side his head.
His shadowy face, mustache and goatee,
Giving stranger aura of being ghostly.

My name is Lucifer now listen me well,
Have a herd of sinners headed to Hell.
The reason to stop in this beer joint,
Have the name of cowboy to ride point.
Final one-way trip shall never return,
Hell-bent cowboy is now going to burn.

Little snockered but only cowboy here,
Chugalug BURP! Hair-lip of foamy beer.

Not exit without kickin and scratchin,

Roll sleeves spit in hands now action.

LUCIFER! I'll tie a knot in your tail.

COWBOY! Your hide pitchfork to impale.

Two enemy warriors combatively enjoin,

Kneecap buried deeply within my groin.
He notched my ear with his dull teeth,
Jerk thumbs from my eyes enjoy relief.
My hair-clumps fly, expose gray roots,
Jackhammer my feet with hobnail boots.
A horned forehead butts me on my chin,

In earnest this fight starts to begin.

Clenched fist rode forearm up my nose,
Imitate a ballet dancer up on my toes.
Broken nose on tilt my blood did drip,
Next to hit his fist HARD with my lip.

Words are filtered thru blood & snuff,

Damn! Lucifer ain't I hurt you enough?
My xylophone ribs bony elbows do play,

Long fingernails gouging my face away.
Lucifer bear-hugs squeeze me to death?
Wilt to smell of my stale beer breath.
My mouth suddenly developing a twitch,

Attempt to swallow a knuckle sandwich.

His pointed toe kicks me in the belly,

Gasping for air as legs turn to jelly.
Slowly wilted to floor position prone,

Pray Dear God for my sins did I atone?
Clear voice speaks and words not mine,

Lot of fun had a DEVIL of a good time.
For now give you a temporary reprieve,
Cowboy unfit for travel need to leave.

Beertender now you can truthfully say,
Today saw cowboy make Satan back away.

REFLECTION
(True)

Black lava beach, sharing sunrise & swirling fog,
Emerging native girl driving all the senses agog;
Classic beauty upon her face emotionally drained,
Immature breasts protrude and salty-tear-stained.

Muscular arms are dangling aimlessly at her side,
Lacks purpose or direction in her halting stride;
Bewildered, eyes downcast, cane dust in her hair,
Red coral necklace; body is skinny, brown & bare.

Recessed navel, proving birth to a mortal nation,
Sans pubes camouflaged vaginal whole of creation;
Her rippling thighs, knobby knees & smooth shins,
Trim ankles, tiny feet passing the field of sins.

Impasse, she didn't curtsy nor did masculine bow,
Consummation dilemma ex post facto & visible now;
Juvenile exposed woman-watching maturity go awry,
Said hello to infamy, now is time to say goodbye.

Into the preceding verses, unknown what you read.
Allow me to interpret for you what's really said;
Dark tragedy, shedding light on a hidden mystery,
12-year-old Samoan samurai was molested sexually.

Red coral necklace, the line to cut off her head,
Machete raised, is what a sugar cane worker said;
Deflower sexual rendezvous consummate prepuberty,
Influencing as child, teen, adult all in secrecy.

Grandmother bore testimony 40 years after attack,
Expose her juvenile body and soul front and back;
Erect, eyes moist, tremble, waiting consequences,
Vote absolution to a special Polynesian princess.

QUESTION?
(True)

While walking thru desert today,
Rattlesnake FREE! Blocks my way.
Side stepping in halting stride,
Rattlesnake slithered alongside.

Now journeyed in same direction,
Both sense an eye of inspection.
Peace, I show no sign of battle,
Not threatened, no snake rattle.

Snake displays grace and beauty,
To kill the creature is my duty?
Samaritanly I leave snake alone,
Hasten my steps & head for home.

Walking alone birthed a thought,
MORTAL ENEMY! Battle not fought.
Seek snake! Life not to prolong!
Absent and alive did I do wrong?

REMEDY

A street cop and today's shift is complete,
Feet are sore from walking my regular beat.
Looking at two swollen ankles in disbelief,
Elevating my legs will provide some relief.

Relaxing comfortably in my reclining chair,
Stared at two bare feet and toes five pair;
45-degree angle toes sequence big to small;
Variety shape, size, width ten toes in all.

Wiggled toes mimicking reeds in the breeze,
Muscle control complete the task with ease;
Horizontal veins on the topside of my feet,
Blood pumping thru try to detect heartbeat.

Between 1st and 2nd joint, top of the toes,
Hair grows. Why? Unknown, if someone knows.
Ten toenails that are growing continuously,
Would I be a sissy if someone pedicured me?

Heels together spreading feet formed a "V",
As sighting through gun sight watched "TV",
Football, who's playing? What is the score?
Favorite sport, eyes close, destiny, snore.

INDIAN LORE?
X
(Narrator-Indian)

A secluded Indian Reservation a responsibility,
1961 as criminal investigator *B.I.A. hired me,
To investigate 11 major crimes on Federal land,
Appointed Police Commissioner by Tribal demand.

Live, work, friendships with Indians did blend,
Three years later my Indian world comes to end.
Original midget Indian, no warrior, a renegade,
Stabbed at too tall heart, downward pull blade.

Cut clear to my manhood my guts trying to ooze,
Midget Indian ran away, high on homemade booze.
Hospitalized thirty-one days and I nearly died,
Full Tribal Council solemnly visits me bedside.

"White man as council we regret a favor to ask,
Midget is sacred to us, don't take him to task.
Neither campfire stories & written Indian lore,
Midget Indian none sighted or recorded before."

With due respect for Indians, midget I forgive,
Mental & physical scar will continue to relive.
Not healed, in writing notify Tribe and B.I.A.,
With mixed emotions, leaving reservation today.

Several years later written message is sent me,
Banish midget, etch tribal shame for posterity.
Failed scalping test! As warrior NOT qualified!
Scalps exhibited center hole or handle outside!

*(B.I.A. BUREAU INDIAN AUTHORSHIP)

CAROL CRANE

Water skiers were many and kept the water rough,
Bass fish were not biting and fishing was tough.
Carol's fishing from within stern of Kim's boat,
No chance of winning if it were put up for vote.

Carol fished shallow; she also fished very deep,
All the fish she caught measured enough to keep.
Observing her standing at the weighing in scale,
She is obviously nervous and turning quite pale.

Carol out fished us she topped the entire field,
Catching the most fish AND heaviest total yield.
7-14-84 Saguaro Lake Tournament for Canyon Bass,
Was won by a lady who showed us plenty of class.

The first lady club member with first place win,
Here is your trophy, tell your story, now begin.
Truthfully, tell us lure, depth and line weight.
What is left hand-inboard-outboard-spinner bait?

COWBOY
(Narrator-Cowboy)

Smell of cologne strongly prevails,
Line dance, Cowboy manicured nails.
Smile for any woman's true delight,
Full set of teeth all pearly white.

Custom made boots expensive & bold,
Massive belt buckle all shiny gold.
Tailored pants, shirt without peer,
Diamond earring in lobe of one ear.

Hairspray hairs never been greased,
A Cowboy hat freshly store creased.
Is articulately citified and bland,
Approach with extended smooth hand.

"Hi beautiful prime lady, howdy do?
Call me Cowboy pleased to meet you.
Capture the moment & dance jointly,
Line dance I see you & you see me."

Dance now over buy her late dinner,
Bedroom before sleep Cowboy SINNER.
She wakes up smiling prior to dawn,
Hungrily reach out, Cowboy is gone.

COWMAN
(Narrator-Cowman)

Rank air of sweat & stock prevails,
Used-Cowman broken fingers & nails.
Hair lip grin snoosh drool a sight,
Five teeth lost, multi a fistfight.

Run-down boots heeled & half soled,
Scratched belt buckle dull and old.
Pants and shirt patches did appear,
Hearing aid sticks out of each ear.

Thinning hair now bag-balm greased,
Stained Cowman hat element creased.
Not schooled, educated by the land,
Approach and extend a callous hand.

"Hey there Woman! Think you're due,
Cowman here to pleasure me and you.
Belly-up hunker-down dance with me,
Dark corner gives us some privacy."

Dance now over buy six pack dinner,
Bedroom before sleep Cowman WINNER.
She wakes up smiling prior to dawn,
Hungrily reach out, Cowman is gone.

(Option)
He wakes up grinning prior to dawn,
Hungrily reach over, woman is gone.

COWBOY/COWMAN
(Narrator-Narrator-Cowboy-Cowman)

Smell of cologne strongly prevails,
Line dance, Cowboy manicured nails.
Rank air of sweat & stock prevails,
Used-Cowman broken fingers & nails.

Smile for any woman's true delight,
Full set of teeth all pearly white.
Hair lip grin snoosh drool a sight,
Five teeth lost, multi a fistfight.

Custom made boots expensive & bold,
Massive belt buckle all shiny gold.
Run-down boots heeled & half soled,
Scratched belt buckle dull and old.

Tailored pants, shirt without peer,
Diamond earring in lobe of one ear.
Pants and shirt patches did appear,
Hearing aid sticks out of each ear.

Hairspray hairs never been greased,
A Cowboy hat freshly store creased.
Thinning hair now bag-balm greased,
Stained Cowman hat element creased.

Is articulately citified and bland,
Approach with extended smooth hand.
Not schooled, educated by the land,
Approach and extend a callous hand.

"Hi beautiful prime lady, howdy do?
Call me Cowboy pleased to meet you.
"Hey there Woman! Think you're due,
Cowman here to pleasure me and you.

Capture the moment & dance jointly,
Line dance I see you & you see me."
Belly-up hunker-down dance with me,
Dark corner gives us some privacy."

Dance now over buy her late dinner,
Bedroom before sleep Cowboy SINNER.
Dance now over buy six pack dinner,
Bedroom before sleep Cowman WINNER.

She wakes up smiling prior to dawn,
Hungrily reach out, Cowboy is gone.
She wakes up smiling prior to dawn,
Hungrily reach out, Cowman is gone.

(Option)
He wakes up grinning prior to dawn,
Hungrily reach over, woman is gone.

ODDS "N" ENDS #1

One and only friend becomes a monk.
Family cat got pregnant by a skunk.

Need glasses and do not have a nose.
Artificial turf on your patio grows.

Come down 11% grade then brakes fail.
Silver Fox fur tarnishes at the tail.

Being embalmed and you have not died.
Loyal and trusty dog commits suicide.

Coach at Notre Dame and not Catholic.
Inherit liquor store, am ex-alcoholic.

Have two pair of shoes, only one sock.
Damn hearing aid only plays punk rock.

Car has four reverse, one forward gear.
World Grand Champion stud, he is queer.

Do not know why yellow snow tastes best.
Spouse fails their sex health card test.

Have a telephone solicitor hang up on you.
House on fire, insurance premium past due.

My name heads everyone's junk mailing list.
Told by Social Security, you do not exist.

Living bra died while you were wearing it.
Learn man flying plane, ex-kamikaze pilot.

Seeing eye dog failed his annual eye exam.
Am main speaker at "American Day" in Iran.

Stung on your nose while smelling a flower.
Well runs dry while you're taking a shower.

Called "Son" by the ugliest ape at the zoo.
Mortician refusing to be of service to you.

Eight Day clock wants 40 hour week, vacation,
Am guest vocalist for annual deaf convention.

Given $10.00 to eat elsewhere by your waiter.
Learn your attack dog is really an alligator.

Mom admits she is virgin, then eternal slumber,
Nice obscene caller says, "sorry wrong number."

Take "Montezuma's revenge" for very fast relief.
Local prostitute asks you, "Where's the beef?"

Bowl goldfish are the strangest I have ever seen,
Brass fish, not gold fish, water's turning green.

Midget outlaw Indian warrior not to be qualified,
Scalps had hole in the middle, or handle outside.

What trouble! Am stopped for traffic misdemeanors,
As ex-wife's lawyer took this cop to the cleaners.

Enjoy the wherewith Mother Nature gave us for play,
Before we're ready Father Time will snatch it away.

CERTIFICATION

Knuckles starting to score,
Knockings to my front door.
Opened door, what do I see?
A Grandson, tall to a knee.

For the first time we meet,
Do come in and have a seat.
He, One year, months three,
Proud you came to visit me.

Move to porch weather mild,
Absorbing the Arizona wild.
Enjoyable trip from Alaska?
Many things want to ask ya.

Talk everything & everyone,
Laughing, joking, have fun.
Chawing from licorice plug,
Pulling from cold milk jug.

Hear a Noise? Quiet, shush,
A creature is moving brush.
Cross our fingers and hope,
Trophy size wild jackalope.

A jackalope is rabbit form,
Head adorns a rack of horn.
Reasons hunters to compete,
One is trophy, one is meat.

Licorice wad grandson spit,
Jackalope took a fatal hit.
Saluted the hunting winner,
Cooked jackalope as dinner.

Think this is a tall story?
There's proof of his glory.
At home, hanging on a wall,
Mounted trophy seen by all.

MY MOUNTAINS
(Narrator-Indian Spirit)

Superstition Mountains our Lord did grace,
Arizona evolution is this enchanted place.
Opulent mountain silhouette to aviary sky,
Geology, fauna and flora kaleidoscope eye.

Chameleon ambience sunrise to sunset days,
Highlight nighttime by stars & moon phase.
Apache Indian Spirits from mountains call,
VIOLATE OUR SACRED GROUND IS DEATH TO ALL!

Lore of Peralta & Lost Dutchman gold mine,
Miners endeavor trespass pristine sublime.
Weavers Needle and Shiprock granite bound,
Landmark clues to gold asleep underground?

Nature, Mountains, Spirits & Mortals rage,
One hundred plus war years they do engage.
Non-Indian hostiles today, Apaches of old,
Kill encroaching prospectors seeking gold.

Legends, mysteries, myths & graves abound,
Modern proof of gold has never been found.
Federal edict! **Mining Superstitions cease,
Exit the mountains and spirits with peace.**

Rugged mountains beauty I awe since child,
Happenstance & danger omit visit the wild.
Now view only as old age stifles my quest,
Mother Nature grandeur! Her pinnacle best.

WINDOWPANE

Looked thru window in awe visibly,
Mother Nature's Christmas scenery.
Pretty Aurora goddess of the dawn,
Lighting ice crystals on the lawn.

Kids & Grandkids outdoors in snow,
Enjoy watching them create a show.
Ice and snow melting enchantingly,
Enhancing to flower, bush or tree.

Birds feeding, singing, surviving,
New snow clouds overhead arriving.
Alerting my family to come inside,
Hug & kiss them & beam with pride.

Gentle snowflakes drift awesomely,
Magnificent beauty! Joyous to see.
Instant eruption of violent storm,
Snowbound, comfortably home, warm.

WINDOWPAIN

Looked thru window in awe visibly,
A family decorated Christmas tree.
Colors, shapes of packages galore,
In tree branches and on the floor.

Candy canes & wreaths on the wall,
Hats, coats, boots drying in hall.
Doorway hangs suspended mistletoe,
Fireplace banked coals, soft glow.

Sounds of laughter, music, family,
Mirrored happy memories escape me.
Cat paws spar Xmas tree ornaments,
Temperature falling, heaven vents.

Gentle snowflakes drift awesomely,
Cold harsh nature kiss of reality.
Instant eruption of violent storm,
No shelter, homeless, to get warm.

WINDOWPANE/WINDOWPAIN

Looked thru window in awe visibly,
Mother Nature's Christmas scenery.
Looked thru window in awe visibly,
A family decorated Christmas tree.

Pretty Aurora goddess of the dawn,
Lighting ice crystals on the lawn.
Colors, shapes of packages galore,
In tree branches and on the floor.

Kids & Grandkids outdoors in snow,
Enjoy watching them create a show.
Candy canes & wreaths on the wall,
Hats, coats, boots drying in hall.

Ice and snow melting enchantingly,
Enhancing to flower, bush or tree.
Doorway hangs suspended mistletoe,
Fireplace banked coals, soft glow.

Birds feeding, singing, surviving,
New snow clouds overhead arriving.
Sounds of laughter, music, family,
Mirrored happy memories escape me.

Alerting my family to come inside,
Hug & kiss them & beam with pride.
Cat paws spar Xmas tree ornaments,
Temperature falling, heaven vents.

Gentle snowflakes drift awesomely,
Magnificent beauty! Joyous to see.
Gentle snowflakes drift awesomely,
Cold harsh nature kiss of reality.

Instant eruption of violent storm,
Snowbound, comfortably home, warm.
Instant eruption of violent storm,
No shelter, homeless, to get warm.

RAPTURE

Reclined au natural on her back,
Prepared for predestined attack;
Widespread legs, cavity exposed,
Give thanks with my eyes closed.

Illumined into a lecherous grin,
Visually caress oily brown skin,
Breast well rounded, body plump,
Legs matured by walk, run, jump.

Body aroma, stimulating delight,
Totally oblivious to her plight;
Proceed, a pleasure to complete,
Carve turkey hen, now let's eat.

YESTERDAY

(WWII 16,112,566 SERVED)
United States of America's flag unfurled,
Championship banner throughout the world.
Military power vulnerable and expendable.
Quit two-face politician NON-COMMENDABLE.

(WWII 670,846 WOUNDED – non fatal)
Presidents, Senators, Congressmen, begin,
Committed to wars we didn't start or win.
WHO foots the **bills**? WHO get the grease?
Rhetoric expounded, IN THE NAME OF PEACE!

(WWII 405,399 DIED)
Fence mender, money lender, sour destiny,
Count the bodies to our place in history.
STOP the wars! STOP burying our war dead!
Be STATESMEN & bury the HATCHETS instead.

TODAY
(Yield vs Victory)

Forthcoming see a "Y" in the road ahead,
Option "V" or silence of our multi-dead.
All *Rainbows coalesce in exact legions,
Some testify, some can't? Fatal reasons.

Battlefield defender of worldly terrain,
Some come home, some destined to remain.
Aircraft and parachutes voiding the sky,
Some kiss the ground & some bid goodbye.

Multi variety of military cast into sea,
Some expelled & some shark shit will be.
Voice of those dying captivate my heart,
Some limbless, shot, blind, blown apart.
Some pray, beg, curse, cry, sing a hymn,
Others, **MAMa. . .Mam . . Ma . m**
Last rites, honorable discharge is read,
In memory I salute those who went ahead.

Politician DEFER! Now campaign military,
Freedom, safety, rights, pride, history,
WE persevere enemy/friend fire, *PTSD.
Politician/military EN GARDE "Y" vs "V".

*RAINBOWS/Aircraft-Mechanized-Watercraft,
Flags-Uniforms-Brothers-Sisters TO ARMS!
*PTSD – Post Traumatic Dress Disorder

*In my generation, there was not much training in
smaller departments, badge #65, we learned by doing.
Verse #5 is another reason.*

#19795

Adult probation officer responsible only to the court,
Pre-sentence investigation then submit written report;
Supervise probationers at the home and in their field,
Make sure that to temptation they don't dare to yield.

Great state of Arizona for murder in the first degree,
And when the 12 jurors can not recommend any leniency;
After polling the jury Judge Lockwood draws in breath,
Have no alternative, the State sentences you to death.

Time of execution is to be set at later time and date,
Probation officer the court orders you re-investigate;
See if anything was overlooked, any rights are denied;
Contact a prisoner who by #19795, shall be identified.

Talk to officers, witnesses and reconstruct the crime;
A confession from #19795, guilty is correct this time.
#19795 shot his wife and then tried to commit suicide,
Doctor saved his life so for his crime could be tried.

Pre-sentence contacted #19795 on almost a daily basis,
Aside from his case we discussed life, its many faces.
Other than male, nothing in common yet bond as friend,
#19795, **"Won't know anybody, will you come to my end?"**

#19795
(continued)

3am 3-7-59 Warden is briefing us at Arizona State Prison,
He informs witnesses what to expect before sun has risen.
8 years varied law enforcement & veteran of World War II,
Was no precedent to prepare me for what I am about to do.

Early morning chill and dampness of courtyard dirt floor,
While standing silently waiting opening gas chamber door,
Any challenge about my valor questionable I will survive,
Result of knowing why #19795 will not leave prison alive.

What is he thinking? Eat any breakfast? Sleep last night?
Making his peace with God? Request or reject? Final rite,
Let's get the show on the road better do it pretty quick,
I think that if you don't hurry up I am going to be sick.

Chamber separated front and rear by thick panes of glass,
Warden, Priest, Doctor in front, witnesses in back amass;
Enter crowded room, will stand through the entire affair,
See anchored to floor a massive special built iron chair.

Built in chest strap, wrist and ankle straps each a pair,
Earpiece outside gas chamber to stethoscope on the chair.
Chute lever outside chamber to under chair crock of acid,
Chamber is empty of people, chair looks tired and placid.

Witnesses discussing if any deterrent value of execution,
Which is better, gas, shooting, hanging or electrocution?
Escorted by guards #19795 entered the chamber hesitantly,
Blankly looked at witnesses, I am sure he did not see me.

Dressed in boxer shorts, undershirt, feet that were bare,
No signs of fear, erect as guards strap him in the chair;
Prison Doctor attached the stethoscope to #19795's chest,
Listens through the earpiece while making one final test.

Doctor, Warden, Priest say goodbye no reprieve to arrive,
Now vacant inside chamber except for apprehensive #19795;
Witnesses watching everything in awe and reverently mute,
Releasing into acid crock cyanide pellets from the chute.

Small puff almost invisible smoke can barely see it rise,
Nostrils flaring, convulsions, his pain I try to surmise:
Four, eight, twelve long minutes, finally bowed his head,
Doctor patiently listens, officially pronounced him dead.

Solemnly pledging self improvement, hard work, education,
Deprived family and myself by law enforcement dedication,
Have thought about this often, now 25 years have gone by,
Executing #19795 did not change the world, neither did I.

> *I have seen death in many forms and have taken human life, #19795 was the most lingering.*

WRITE
(Poetaster)

Pen myself another rhyme today,
Satisfied that I did it my way.
Utilizing only God given tools,
Never introduced to poet rules.

Style of others not to measure,
To **write** gives myself pleasure.
Perceive me either good or bad,
Bare my **LIFE** or **THOUGHTS** I had.

EXHIBIT! When any poem is done,
Some fit reader more than some.
Inconsequential of poems worth,
Admit paternity to their birth.

Perchance my **writings** you read?
If I fail to fulfill your need,
Challenged now to a little fun,
Equal my style & **write** your #1.

EVOLUTION

Night opts day reveal universal surprise,
Showcase God, Mother Nature's enterprise;
Nurtured balance through formative years,
Awing nightly, fulgent celestial spheres.

Moon full, different quarters, concealed,
Energy, heat, light, sun quotas to yield,
Flame, molten lava is internal consuming,
Known to unknown unquestionably assuming.

'Clouds and blue skies are earthly arched,
Hostile flora· enchanting deserts parched;
Varied forestry with cornucopia greenery,
Generate oxygen, shelter, food & scenery.

Four seasons, wrath or beauty undisputed,
Oceans, rivers, lakes, waters unpolluted.
Metals, roots, minerals buried treasures;
Hear, taste, smell, see & feel pleasures.

Intermittent raindrops drizzle to abound,
Rainbow iridescence phenomenally astound;
Lightning exhibited to thunder's applaud,
Sunset artfully reclining to Land of Nod.

Environmental organisms forfeit defiance,
With Adam and Eve's seed formed alliance;
Entrusted man, preserve the world, enjoy,
More powerful than God! Watch us destroy.

MIRAGE

Boat, trailer, truck from garage,
Routed to fishing Arizona mirage.
Heat rises, relate H2O confusion,
Reflected light optical illusion.

Locate spot where sun does glare,
Non-critical & can park anywhere.
Effortless fishing & has my vote,
I am not required to launch boat.

From truck to boat is transition,
Dry & make ice chest acquisition.
I loft first beer to sun & toast,
Admire mirage created by my host.

Flaunt beer no mirage cop patrol,
Minus truck driver I can't troll.
Removed shirt & reject life vest,
Sun feels good upon a bare chest.

Opened beer in my hand very cold,
Fishing story is near to be told.
Mirage fishery of wildlife image,
By rod, reel & beer will pillage.

Open second beer in haste & zeal,
Pick up my favorite rod and reel.
To my line attach my secret bait,
Balance of 1st 6-pac, annihilate.

Kidneys prevail to empty bladder,
Replacing beers makes me gladder.
I cast out with my favorite lure,
Guzzle many a beer at my leisure.

With magic water wand, flood bug,
A beer in either hand chug-a-lug.
Upon forehead perspiration beads,
Coldest beer fulfilling my needs.

My kidneys again I need to flush,
Sip fresh beer & empty can crush.
Aware dehydration by Arizona sun,
A beer, fortify with another one.

P'nut peed my name into dry sand,
Beer is clinched in my free hand.
Getting sunburn, my senses alert,
Procure beer while reclaim shirt.

Cheer my doctor & his dedication,
Chase with beer, take medication.
Pee for altitude, record I claim,
Sans beer, fish'n ain't the same.

Darn mosquitos do hurtfully bite,
Ingested beer adds to my delight.
Upset stomach is getting seasick,
Grab can of beer & down it quick.

Open fly, extract shirttail, pee,
Cold beer, hot H20 cascades knee.
Right shoe dry & left shoe soggy,
Drink beer faster getting groggy.

Retrieve my lure with rod & reel,
No stink, no fish to gut or peel.
Fall overboard into my truck bed,
Pillowed sleeping bag at my head.

Arizona sunset! Best of anywhere,
Bats foraging cooling desert air.
Man-in-moon & stars jointly wink,
Own bloodshot eyes heavily blink.

Sea horse herd dreamingly whinny,
Mermaid choir beautiful and many.
Saguaro octopus arms do windmill,
Boat porpoises as mirage unstill.

Various whale pods cloud the sky,
Mirage sharks circle dorsal high.
Drift past alligator lined banks,
Unwary diver, bubbling air tanks.

Ambergris odor prior to perfumer,
King Neptune abdication is rumor.
Periscoped by tortoise submarine,
Intermission! Pee & resume dream.

POTLUCK
(Narrator-Chairperson-Senator)

Park my car, Senior Center,
Take picnic basket & enter.
Strangers, friends in hall,
Greet, visit with them all.

A food contribution inside,
Picnic basket births pride.
Serving table asks finesse,
Consorting conspicuousness.

Assigned table where I eat,
Set table then take a seat.
Time's moot before we fare,
Place settings, to compare.

Chairperson spoons a glass,
I request silence en masse,
Now before we do our thing,
National Anthem, must sing.

Think culinary art to brag,
Pledge allegiance, no flag?
Pause to breaking of bread,
Food blessing must be said.

Senator, **"Bless in Luster!"**
Presenting us a filibuster.
Recognize your need to eat,
9 member committee do meet.

Before chow we can partake,
Many announcements to make.
Prior to slurping the soup,
New members into our group?

Feast time is getting near,
Any guests we hope to hear?
Await temples of cookeries,
Birthdays? & Anniversaries?

Will banquet soon, not yet,
Engaged?, Wedded?, New Pet?
Remind before meal to seek,
Belchfest every other week.

Ennie, Meanie, Minee & Moe,
Selected first table to go.
Sorry but I'm so perplexed,
I don't know who goes next.

Choice I choose to be last,
Sideshow before our repast.
As member, played my hunch,
5pm potluck, did 2pm lunch.

MOTHER-TRUCKER
(West Trucker-West Beertender-South Beertender-East Trucker)

Proud Indian Squaw & provider to our house,
Got my own tribe of kids & a loving spouse.
I don't pop pills, inject drugs, smoke pot,
But do frequent country-western bars a lot.
As mother-trucker 18 wheeling any locality,
Ask people to play cowboy & Indian with me.
Stand at the bar & in a voice loud & clear,
I prefer bourbon-shots and chase with beer.
I'm a big, bad, mother-trucker of the west,
Drinking, fighting as trucking-Indian BEST!
On the prod seek anyone who wants to fight,
Loser buys winner drinks the rest of night.

Indian mother-trucker eyeballing the scene,
Customers act invisible, exit, use latrine.
86 mother-trucker or eat beertender sixgun!
Leave before I either dial mortuary or 911.

Peace beertender, time I hit trucking road,
Trucking to Florida to empty trucking load.
Three days later & trucking the deep south,
I'm thirsty, spitting cotton from my mouth.
Down the trucking road see a sign, it read,
Rednek hospitality roadhouse 8 miles ahead.
Not required to wear either shirt or shoes,
We accept food stamps for tobacco or booze.
Need of recreation, will roadhouse tonight,
Challenging! Southern hospitality or fight?
Stand at the bar & in a voice loud & clear,
I prefer bourbon-shots and chase with beer.
I'm a big, bad, mother-trucker of the west,
Drinking, fighting as trucking-Indian BEST!

I'm rednek beertender & listen to my voice,
Beard you mother-trucker, gave us a choice.
Ladies-night, southern belles rednek tough,
Injun mother-trucker I accept tough enough.
Rednek hospitality honorary southern belle,
New rednek! Free drinks, stand to the well.
Instant rednek country cousins quite a few,
If trouble tonight share SHOWTIME with you.

Stand at the bar & in a voice loud & clear,
I prefer bourbon-shots and chase with beer.
I'm a big, bad, mother-trucker of the east,
Drinking, fighting as trucking-queen BEAST!

I'm rednek beertender & listen to my voice,
Beard you mother-trucker, gave us a choice.
Ladies-night, southern belles rednek tough,
Beast mother-trucker I call candyass bluff!
Commitment made I now give a choice to you,
SBOWTIME opponent choose either one of two.
Rednek hospitality opponent, TABASCO CAJUN!
She de-fangs beasts I dial mortuary or 911.
Rednek hospitality opponent, WARPATH INJUN!
She scalps beasts & I dial mortuary or 911.

Peace beertender, time I hit trucking road,
Trucking to Arizona to empty trucking load.

*86-Beertender terminology for-cut off-
no more service-get out-etc.

OBSERVATION

Roll down windows it is getting very hot,
Sitting here in our car in a parking lot.
Waiting for wife can sometimes be a bore;
Probably buying out the department store.

Kwitcherbellyakin and take a look around,
Might be surprising, what could be found.
Avoid getting hurt or cause verbal abuse;
Pedestrian crosswalk, there for your use.

Different automobiles brand new to relic;
People's face range from mean to angelic.
Children skateboarding upon the sidewalk,
Adults at a baby carriage talk baby talk.

Sunglasses in various shape, color, size,
In purses, shirt pockets and shaded eyes.
Licenses from all of the snowbird states,
In handicapped parking commercial plates.

Height of people vary from short to tall,
Sandwiching bodies large, medium & small.
Shopping flyers under a windshield wiper,
Chrome shining, showpiece, a clean biker.

Hear the siren as busy ambulance goes by;
A beautiful day goes someone have to die?
Speaking of dying I am dying for a smoke;
Quit, still crave, lung cancer's no joke.

Birds into the landscaping, into the air,
Released their droppings, without a care.
A family sleeping in their car destitute;
Girl working the street local prostitute.

A beautiful build on the great big black,
Bet he is some All American running back.
Vehicles parking in the no parking zones,
Kid checking coin slot in the pay phones.

Hoods rising on two autos, consecutively,
Jumper cable reaching battery to battery.
Designer jeans sure do show off the buns,
Do security police qualify to carry guns?

My wife, a brand new dress she's wearing,
I will help with packages she's carrying.
For purchases she used my charge account;
Number of packages, dread a total amount.

As we walk back to the car I really know,
A parking lot is but one continuous show.
I'll return as there is much more to see,
Why is that guy in his car looking at me?

CONSCIENCE
(Person-Voice)

From mirror hanging on the wall,
My name, voice does softly call.
"Time crucial, please come here,
Confront mirror, I will appear."

Face to face prolonging mystery,
Mirrored image is talking to me,
Identity of voice I have a clue.
"Mortal I convey message to you.

Your attention hope I captivate,
Before knocking the Pearly Gate.
My gatekeeper is almighty judge,
Sans approval gate not to budge.

He is ultimate power to forgive,
You have a power while you live.
Forgive! Abandon hatred & anger,
Now communicate without languor!

Confront, telegram, mail, phone,
Forgiving omits mandate condone.
Some unlocated or passed beyond?
To them premiere way to respond.

Close eyes, bow head & now pray.
Gatekeeper records & will relay.
Burden now light by half a load,
Edict! Complete absolution mode.

In appearance you are very nice,
Bet water unwalkable unless ice.
Double-edge-sword-cliche comply!
Seek forgiven same rules apply."

TO BE CONTINUED

Hospital 1974, rang nurse bedside,
I predicted then immediately died.
St. Pete cheerfully, "To Hell go."
Devil fearfully, "NO-NO, HELL NO!"

Both rejected, return me to earth,
Another chance now prove my worth.
Expose emotion by fact or fantasy,
Human, celestial & natures family.

Cleansing my soul I shall confess,
Poets' secrets for you to possess.
Final lifetime experiences? I bet,
Now 1994, not yet friend, not yet.

MEMORIES
(Me-Art)

I had a very generous Dad,
He shared anything he had.
Nephew Arthur was in need,
Standing horse can't feed.

Opts boarding for his use,
Our pasture run him loose.
True name of horse is Rex,
Didn't mention is complex.

6th grader wanted to ride,
Bridle horse jump astride.
Determine if have a brain,
Left, right, do neck rein.

Giddyup & whoa & now back,
Rein tight restrain slack.
His mouth is hard I admit,
Wish we had a Spanish bit.

Time to have a little fun,
Come on horse can you run?
Running fast straight out,
Hi-Yo-Silver loudly shout.

Rex captured bit in mouth,
Turns, I go west he south.
Soared until I hug a tree,
Rex did neigh grandiosely!

Remove bridle return home,
Horse is pastured to roam.
53 years later as we walk,
Cousins reminisced & talk.

Once owned horse name Rex,
Son-of-a-gun very complex.
Freshly remember one time,
Art's story, silence mine.

DAVID

Mesmerized daughter whom we call Jane,
Married her & changed her maiden name.
Arizona wed and Alaska homeward bound,
Union later results children profound.

Daughter Jane's welfare? Fear omitted,
Respect, love you as family committed.
Stated many times in a positive voice,
Could not have improved Jane's choice.

Favorite Son-in-Law lo the many years,
Loyal, proud, David Michael Breshears.
Very pleased the way families blended,
Name you as my #4 son if not offended.

ODDS "N" ENDS #2

See role models change into reverse,
Woman's the doctor, man's the nurse.

Child says, Mama's naked, see no nuts,
Does not have wee-wee, just two butts.

Danced my bride the night we were wed,
Danced and danced, bedroom had no bed.

Electric chair blew fuse, now sit back,
Waiting release, die of a heart attack.

Corner the gold market by being so bold,
Next day Russians invent synthetic gold.

Mother is killed by auto, very sad tale,
Autopsy report to show, mother was male.

Kissing your girl and how she throws fit.
Lost her chewing gum, you are chewing it.

Seek favorite relative, if put to a test,
Of all my relations, I love sex the best.

Announcement releasing to all the media,
KKK leader, dying of Sickle Cell anemia.

Eating breakfast, but there is one catch,
Second soft-boiled egg, started to hatch.

Children called you names, out of respect.
As adults names the same but call collect.

Need white flag to surrender, will be dead,
Don't wear underwear and my uniform is red.

Purchase 50 inch color TV, blow your mind.
Sit back, enjoy, your eyes go color blind.

Given World Series tickets, now hog heaven,
Games finish in six, tickets are for seven.

Kiss her nose, toes & everything in between,
Next day, are on adult channel movie screen.

Sex change operation did not solve troubles?
Learn did not need it to play mixed doubles.

Dating beautiful girl, could not resist her,
Now mistress, learn she is long-lost sister.

Black Americans who return to Africa to farm,
Penalty, leave with a Redneck under each arm.

Quick inform your loving non-smoking spouse,
I quit smoking, everywhere but in our house.

Date says, if you kiss me please do it quick,
I have a strong stomach but you make me sick.

Caught worlds' record fish as you had desired,
Throw back, yesterday fishing license expired.

Double mixed emotion causing your heart attack,
Mother-in-law drove off cliff in your Cadillac.

MASON'S LETTER
8-17-94

My Brother, present to you,
A child who you never knew.
Daughter, proud son's seed,
Masonic help she does need.

6/30/88 this child is born,
Pretty baby health forlorn.
Born deaf & signs to speak,
Kidneys tiny and legs weak.

Custom shoes fit sore feet,
Her health care incomplete.
Problem cause not designed,
Circumstance, finance bind.

Masons I am requesting you,
Any hospital pass her thru.
Doctors will be made aware,
Total of damage can repair.

Endowed by Masonic embrace,
Modern miracles take place.
On completion of your task,
Thanks for doing as we ask.

Walter Phillip Jordan, Jr.
Mason, Past Patron O.E.S.,
32nd Degree Scottish Rite,
Demolay Chapter Dad, Past,
Rainbow Grand Cross of Colors.
(Kingman, AZ #22 F. & A.M. - NO REPLY)

KEEPSAKE
(True)

Sitting upon entrance of driveway,
Await my return from school today.
A black cocker spaniel named Inky,
Killed by car and driver so kinky.

Tire tracks from road to shoulder,
Purposely stop Inky growing older.
Devastated by the sight before me,
Driver ran over Inky deliberately.

Raised from knees by gentle voice,
"Youngster, time to make a choice.
Dog's dead, now have things to do,
Family not home? I will help you."

Lip quivered as I try to be brave,
I hug Inky, stranger digs a grave.
Not professional, it was our best,
Inky wrapped in my shirt, at rest.

The tragedy of Inky now forgotten,
Also is anger to driver so rotten.
Fondly remembering kind Samaritan,
Compassion for boy, not yet a man.

DAH-GO-TEH
(Greeting)

Enforce Federal Law for 11 major crimes,
Plus Tribe Ordinances in boundary lines.
Tribal Police Commissioner, job *B.I.A.,
Chairman Clarence Wesley, greeted today.

Walter Phillip Jordan Jr. my given name,
Phil or *Enah Ahsitiny I'll answer same.
San Carlos Arizona, 12-31-61 to 9-26-64,
Was housed opposite Apache Tribal Store.

Political Apaches fail to intimidate me,
Undaunted Peace Pipe offer share coffee.
Tomahawk no threat, I never carry a gun,
Had favorites but try to equal everyone.

Tribal Police, Judge & Clerks ply pride,
Dedicated, professionally protect tribe.
Hire *Arminda Antonio first policewoman.
All respected in counties Gila & Graham,

Felonies, misdemeanors fill tribal jail,
From Tribal Chairman to drunks opt bail.
Appreciable reduction in recorded crime,
To all children Law & Order donate time.

Sponsor wrestling profits for pool hall,
Halloween parties and teams of baseball.
Children gifts, police freely give away,
Aided by trading stamps get from *B.I.A.

Gather hundreds of books, start library,
Reservation more stable & less contrary.
Condensed memories, dynamite, fry bread,
Cattle sale, Coolidge Dam, soda pop red.

Safford, Globe, soup or dumplings acorn,
Asbestos, drown old San Carlos, newborn.
Old Town Rice, wine, Peridot, drum ring,
Horse, Grandmother Bear, tortilla, sing.

Valley of Sick, cradleboard, wild honey,
Bootlegger, sardines, tribal gold money.

Surplus food, new & old courtroom, jail,
Wickieup, FlagDay, Apache fiddle, quail.

Camp dress, sweat baths, *Tulapi, piñon,
Corn, Point of Pines, Tufa & blue stone.
Black River, lake, tank, pond or stream,
Fish, eagle feather, Medicine Men dream.

Burden basket, gourd, sawmill, owl hoot,
Fire fighting, basketball, turkey shoot.
Santa helper, wildlife, sunflower seeds,
Pitchpine jug, bonfire, crafts of beads.

Crown Dancers, railroad track, buckskin,
Devil-Claw, tamale, copy non Indian sin.
Mistletoe, wake, pinto or mesquite bean.
Bylas, cowboy, jerky, Miss Indian Queen.

Shady Cottonwood tree brags versatility,
Firewood, carved root or hanging effigy.
Multifaith, coming out dance or sunrise.
Tribal Sovereignty, sunset say goodbyes.

Dennis Nelson, Burnett Cassa host party,
Marvin Mull Chairman, tribe gifts to me.
Two cultures divided by pigment of skin,
Blended as brothers by heartbeat within.

Mohave County Chief Deputy, many a year,
A contingent from San Carlos did appear.
Vice Chairman Harrison Porter and 4 men,
Asks me to be Police commissioner again.

Elected sheriff, four-year term did win,
Chairman Kitch Phillips, requests again.
Humbled and deep in my heart I do yearn,
Anti B.I.A. reasons prejudice my return.

Life's journeys so many places did roam,
Phoenix born, San Carlos felt like home.
To San Carlos Apaches do make a request,
Buried at San Carlos my remains to rest.

A Council vote if given your permission,
Advise location and rules for admission.
Or: Cremated ash on invisible wings fly,
*Enah Cheedin seek any who vote to deny.

*B.I.A. - Bureau of Indian Affairs.
*Enah Ahsitiny - White Policeman
*Arminda Antonio - May be the first full blood female
Indian police officer certified in Arizona.
At the time I wrote this there was
no written Apache language.

*Tulapi - home brew of fermented corn
*Enah Cheedin - white ghost

CAMPGROUND
(Narrator-**Hostess**)

R-Ving a desolate Texas Highway,
Seeking park for overnight stay.
Negative planning was not smart,
City 127 miles by mileage chart.

Talking to myself short of cuss,
Driving time two hours and plus.
Eyelids heavy starting to blink,
Bolding tank's alarmingly stink.

Red dash light campaign for **oil**,
Thirsting radiator soon to boil.
Gas gauges show dangerously low,
Hot crosswind continues to blow.

On windshield gutty bugs cement,
Frustration adds drivers lament.
BARK! Unlit sign along the road,
Exit ramp for OUR CAMPING ABODE.

8 miles anything you're wishing,
Campground, pool, lake, fishing.
Boating, restaurant, cable TV's,
Ice, laundry, chapel, groceries.

Bolding tank & garbage can dump,
Both butane & gasoline can pump.
Volleyball, 18 hole golf course,
Tennis and hospitality in force.

Arrived at gate long after dark,
Sign, **WELCOME! & please do park.**
Please be quiet don't honk horn,
Can register & tour on the morn.

Door knock! Roll out of my bunk,
Voice, "Coffee & donuts to dunk.
OUR CAMPING ABODE agent of host,
Bid welcome & facilities boast!"

I open door & silhouette to sun,
Smiling lady with clothing NONE!
Dumb & numb, brain sounds alarm,
DEEP Do-Do! Picked a NEKED farm.

Words filtering thru silly grin,
Thanks, will you please come in?
She sofa sat upon towel carried,
Define camp & recreation varied.

Enamored by her pair of ?? eyes,
My compulsion is total surprise.
Author cheque, day of September?
What is the cost to be a member?

MUSTANG

Few mustangs still running wild in the far southwest;
In my area one horse stands alone above all the rest.
Black stocking hoofs, mane, tail, hair a grayish tan;
Brown eyes that state, never be ridden by mortal man!

Conformation as if sculpted by critical cowboy hands,
Nomad of the desert, guardian of several small bands.
Buckskin stallion leading a valiant, decreasing herd;
Ownership any cowboy dreamingly would have preferred.

Listen with trust to Mother Nature guiding their way,
Surviving an environment that's challenging each day.
Victimized by all the elements created by God or man;
Beautiful creatures living each day the best any can.

WATERHOLE

Drawn by team of horses wagon is being pulled uphill,
En route to a drying waterhole & fix broken windmill.
Wagon loaded with repair materials, tarp, horse feed,
Bedroll, food, water, tools, a commitment to succeed.

Top the hill, rest horses and admire tranquil valley;
Looking down on waterhole destination, what do I see?
Appears to be a very large animal bogged down in mud,
Binoculars to reveal it is big mustang buckskin stud.

Descend steep rocky hill & bounce on hard wagon seat,
Wheel brake on, smell wood scorched by friction heat.
Horses forefeet bracing haunches dragging the ground,
Dust swirling as all reach the bottom safe and sound.

Arrive at the waterhole full of excited anticipation;
Start looking around, evaluate the present situation.
Various animal tracks around waterhole in drying mud;
Mustang in center captive & covered with dried blood.

Unshod hoof & double cougar tracks both tell a story,
Mustang is fighting for his life, not just for glory.
Mustang enters into muddy water not to fight anymore,
Cougars unable to follow dare mustang to come ashore.

Cougars continue vigil until prey attracts them away,
Mustang by this time is mired in mud, he has to stay.

Weak from wounds, hunger, fighting to raise his head,
I must hurry to help mustang or he will soon be dead.

Collect brush, cacti skeletons, limbs, any dead wood;
Place it in mud making a path to where mustang stood.
Mustang squeals, snorts, bravely commits for a fight,
weak quivering body yielded to mud binding him tight.

Tried slipping nosebag with grain over drooping head,
But this rank son-of-a-bitch tried eating me instead.
Daily buckets, grain to relieve hunger, water thirst,
Prior mustangs extraction must regain strength first.

Burning hot running-iron on wounds need to cauterize,
Daub axle grease on all wounds to keep off the flies.
While iron's hot it would be easy to put on my brand;
No, didn't earn it, mustang wasn't caught by my hand.

When next we meet championship is between him and me,
Able to brand or not determines who claims a victory.
Repair windmill, maintain team of horses, mend fence,
Till mustang & strong enough to restore independence.

Awake! It's morning, yawn, stretch, slept very sound;
Look to mustang GONE! Search quick, he must be found.
Mixed emotions of dismay, happiness, fear did remain,
Till observe mustang at wagon nuzzling sack of grain.

Whinnies as if to say, **"Contents I definitely savor"**.
Admiring mustang, notice left front leg he did favor.
Circle mustang so I can see leg from different angle;
Compound fracture, from immobile leg hoof did dangle.

Stouthearted bastard one problem he could not stifle;
Numb in disbelief, shuffled to wagon for 30-30 rifle.
Tip Stetson in salute, aim surely at head of mustang;
Looking each other in the eyes squeeze trigger, BANG!

DELINQUENCY

**Few parents are very fast at casting blame,
If any child in trouble carries their name.
They make repeated excuses for their child,
And blaming society for making him be wild.**

Since the day of birth we raised him right,
Give him everything and show him the light.
Can not be a very bad child you must agree,
Could not be bad, because he belongs to me.

Our son's not doing well in private school,
But want you to know my son isn't any fool.
Lots of pressure, has not been at his best;
The poor little child doesn't get his rest.

It's hard for us to supervise him you know;
My husband and I are both always on the go.
We only go out and party every other night;
Do not take our son, are raising him right.

We always leave him money so he can go eat.
And I warn him never to play in the street.
While we are gone he can study or watch TV,
Or has numerous friends that he can go see.

I know that our son runs with a good crowd;
They are great kids neither rowdy nor loud.
No we have not met any of them but you see,
Know all about them because my son told me.

Husband plays his regular Sunday golf game,
Except for times that its happened to rain.
Sunday when it rains go to Church and pray,
Rain disappear daddy waiting to golf today.

Husband's exhausted when days work is done;
Daddy is too worn out to play with his son.
Do everything that we can for our dear boy,
Buy him anything that will provide him joy.

You are wrong I know that no child of mine;
Would ever dare consider to commit a crime.
What ever you say can not ever convince me,
That any parents help to cause delinquency.

What's all this crap you are trying to say?
For his parents crime our child has to pay?
That as the child's parents we have failed?
Few parents, not children should be jailed!

SATURDAY NIGHT
(Cowboy-Female Dancer)

Dismounting horse, spit out dust from the trail,
Loosen cinch, air horses back & hitch to a rail.
Arizona cowboy in town to celebrate his divorce,
Party Time! So women let nature take its course.

Enter"END OF TRAIL", Honky tonk of great renown,
Drink, gamble, dance with women from every town.
Juke Box, mechanical bull, radio on police-band,
TV, pool tables, weekends live music in command.

Rented band starts playing lively western swing,
Microphone in hand, country lady begins to sing.
Table near dance floor set beer pitcher & glass,
Grab two chairs, vacancy is for any lonely lass.

Enjoying sight, sound, few beers & feeling fine,
Dancers circle up two old fogeys way past prime.
She's tall, overweight, hair dyed brilliant red,
He's short, skinny, not a hair on his bald head.

In unison dancers clap hands raising tempo beat,
Old man joins me and my empty chair he did seat.
Regular customers know she's anticipating dance,
Alone, relaxed, eyes close, simulating a trance.

Graceful snake-like movements, take exotic turn!
Like on a fish hook wiggles like a big fat worm!
Shimmies & body quivers like gelatin on a plate!
Like two bear cubs in burlap-bag boobs activate!

Now cartwheel, back-bend then into a hand-stand!
Exposed all of her assets and the promised land.
Sounds of applause, whistles and calls for more,
Happy people now joining her on the dance floor.

She eyeballs me lovingly and gives knowing wink,
Raised my glass in salute & chug-a-lug my drink.
She continues dancing and later she does appear,
Hugs me, kisses my cheek and whispers in my ear.

Does not look like much, this dried up old fart,
He has hidden qualities that stimulate my heart.
He & I exit for Vegas to conjugal, tell your Pa,
Reunion upon return. Reply, he #8 or #9 Grandma?

FROM GRANDPA

1947-C46/47/82 Aircraft-T7 Chutes.
Grandpa-Glider & Jump Wings-Boots.
2012-C117/130 Aircraft-T11 Chutes.
Loren-*EIB Badge-Jump Wings-Boots.

2013 Christmas, now hear my voice.
82nd Airborne, <u>our</u> unit of choice.
Religious faith blends militarily,
Stout of heart be best you can be.

Peace salvation is a wishful plan,
Deploy for resolve in Afghanistan.
Training? Mind, heart & body test,
You or your opponent, who is best?

Proudly help a country to be free,
Journal tour & everything you see.
Return schedule now 9 months away,
God, bless our military & the USA.

*EIB-Expert Infantry Badge

P.L.F.
(Parachute Landing Fall)
(Koch pronounced Cook)

Harry Koch used an expression, GUNG HO,
As civilian he wanted to add, GERONIMO,
Ask ex-paratrooper Jordan he will know,
Satisfaction to holler, LOOK OUT BELOW!

Desire is so great he couldn't bear it,
Find jump instructor named Chuck Merit.
Soon located Chuck and started pumping,
Many questions about parachute jumping.

Fraternally stare in Harry Koch's eyes,
Observe jumping fever continue to rise;
Excitement concealed desperately tried,
To keep it from showing on the outside.

Finally when he can't stand it anymore,
Went up in the plane then out the door,
Waited; cussed, prayed, he kept hoping,
Nothing would delay this chute opening.

Then he felt the painful opening shock,
Like being hit in head with a big rock,
Instead of enjoying and looking around,
Could not keep his eyes off the ground.

Asked numerous questions, had one left,
Forgot to ask anyone how to do a P.L.F.
Harry found himself in a Hell of a fix,
Landed as if he carried tons of bricks.

Final end result one broken right foot,
Leatherneck, Deputy Sheriff Harry Koch,
Jumping takes lot of training you know,
Ya only had guts Harry, DAMN GOOD SHOW!

WING SUNG LEE

Teenager to instant adult when father died,
Official cause of death? Committed suicide.
By his death and our customs it's up to me,
As elder son I'm head of my Chinese family.

In Alaska was employed by cousin Sonny Gar,
Restaurant job till lawful age to tend bar.
This is where I met Anglo woman Peggy Bart,
And this is where all my trouble did start.

Peggy and I traveled about as man and wife,
And I didn't know about her dark past life.
Was having too much pleasure for me to see,
What trouble is ahead for Peggy and for me.

We came to Arizona very soon we were broke,
Arrested after second bogus check we wrote.
City officers booked us in the county jail,
Stayed two months could not raise the bail.

We plead guilty and the judge set the time,
That we will be sentenced for felony crime.
My sentence probation, ordered restitution,
Peggy sentenced one-year penal institution.

Second judgment that I'm committed to face,
Are family and the culture of Chinese race.
Family admonishes me & I acknowledge blame,
Edict! Go forth & restore honor to my name.

My cowardly heart concealed one great fear,
That everyone would not see me shed a tear.
I desire all of the people to watch me cry,
Tell them I am miserable and wanted to die.

Finally after self-pity I begin to realize,
People did not really desire to sympathize.
With any person who whiningly admit defeat,
Gutless & not stand up on his own two feet.

Into each life is some rain that must fall,
Just wait; I only get one drop, not it all.
By looking around at other people I'll see,
Others have problems besides Wing Sung Lee.

Unbow your head and purposely hold it high,
Remorsely look anybody straight in the eye,
Hopeful any critics will eventually relent,
When people see that I am trying to repent.

Most definitely it will be very worthwhile,
Observing masks of scorn change to a smile.
A successful quest will showcase my accord,
Aided with help and guidance from Our Lord.

Now as Chinese American with a normal life,
Dream of multi children with a future wife.
I can counsel children for their happiness,
Because I have learned from my foolishness.

I went to Washington to see Uncle Moon Lee,
I asked him to provide a steady job for me.
It's new to me & not my usual line of work,
I am very happy in my job as grocery clerk.

Receiving a letter from my mother overseas,
She arranged for me a bride who is Chinese.
Depart for China via plane before too long,
Wed and honeymoon in the city of Hong Kong.

Knowing by oriental custom I can lose face,
When my family tells all about my disgrace.
New bride accepted me with my tainted past,
I'll do my best to make this marriage last.

Kneeling nightly at my bedside I will pray,
Ask, for forgiveness prior my judgment day.
Have done my best, maybe a little bit more,
I hope end results have evened up my score.

Will be void of fear when it is time to go,
With a pure heart, my mind will truly know.
Regardless of human race, color, and creed,
God may bestow award for any repented deed.

(Revision of original poem; also names, dates &
places changed. Otherwise, correct poem written
from observation and letters received from this man.
Original poem mailed, no reply, correspondence ceased.)

FISHERMEN

Driving to Lake Hodges prior to the morning sun,
4am 8-20-88 munching breakfast on croissant bun;
Two adults sleepily discussed bass fishing lore;
Enough fishing tackle, could open a small store.

It is first light as we arrive at lake property,
Park car, rent boat at dock and pay fishing fee;
Neither has ever fished on this particular lake,
Study topographical map to minimize any mistake.

As for piloting the boat, each took equal turns,
Fish all day from top water to artificial worms;
Three non keeper bass plus! One freshwater clam,
A non-productive fishing trip? Well let us exam.

Professional overrun (bird nest) from each reel,
Airborne crank bait attacked! By bird with zeal,
Quantity of plastic worms we grudgingly donated,
Causing rocky bottom to be continuously rotated.

Few fishermen running & gunning the entire lake,
Bullied fishing spots then left a bouncing wake,
Adult woman shouts, **"you ain't seen nothing yet"**
In her boat, rotates body, nude, glistening wet.

Sans drakes female mallards feeding in the rush,
Scattered quail on the bank call from the brush.
Cantaloupe, chicken sandwich, candy bar & punch.
Drift, personal conversation while eating lunch.

Sharing stories of dreams, heartaches, employed,
Family, jokes, anger, mistakes & things enjoyed;
Partner tells love for wife, two sons, daughter,
Egrets marvel with us as three deer drink water.

Arms, face sting compliments of the sun's wrath,
Come off the lake contented and in need of bath;
One fisherman is happy old man age of sixty one,
FIRST time, fishing with his 29 year old #3 son.

DON'T READ THIS UNTIL YOU READ THE POEM
A man in his late 60's early 70's was slowly overtaking
us in his boat. Lying on the bow is an adult female in shorts
& t-shirt, face down with her arms cradling her head. I told
the driver that he had a good-looking hood ornament & he

repeated to the woman what I said. The woman while standing up stated, "You ain't seen nothing yet!" She rotated, flexed, flaunted her young, well endowed, sweaty body then without another word returned to her prior position. I could not put all of this in the short poem so in describing her body I wrote "full." Now in a fishing story you have to have a "fish story." I replaced the word "full" with the word "nude." Everything but the word "nude" actually took place on our fishing trip. Brad, was she blonde, brunette, or redhead?
What was the color of her shorts & t-shirt?
Dad (Brad Rexton Jordan 12-20-58 to 4-13-17-Friday the 13th)

RATED-X?

Since ways and days of old,
Law enforcement tales told.
Some good and some are bad,
Some happy alas mostly sad.

Old England, before police,
Cop a Conservator of Peace.
Cops arrest, jail to court,
Document by written report.

Dot all I's & cross each T,
For lawyer, judge and jury.
Cops regular duty to shirk.
Handcuffed with paper work.

Abbreviation non-demeaning,
Intent is part of gleaning.
Originate only as shortcut,
Call abbreviated word smut.

Abbreviation birthing word,
Controversy when now heard.
Crime identity acknowledge.
Fornication Unlawful Carnal Knowledge.

Anytime this word is heard,
Reflect how birth occurred.
Word defined & now stained,
By persons NOT cop trained.

CAT BURGULAR?

Startled awake one **dark night**,
Too angry to recognize fright,
From my nightstand grab a gun,
Feet hit the floor on the run.

Run through the **darkened hall**,
Shoulder scraping on the wall,
Instant eerie screeching wail!
Realize, stepping on cat tail.

Jolted back into sane reality,
Painfully hit table with knee.
Hop hop, floor lamp, stub toe,
Entwined, floor bound both go.

Angry! **"Dirty son-of-a-witch"**,
Lamp in hand, turn the switch.
See a blank space on the wall,
Prize cuckoo clock had a fall.

Clock lies broken what a mess,
Cause of **anger**, pain & stress.
Deduction, the ruckus I heard,
Siamese cat after cuckoo bird.

My hand still holding the gun,
Will reduce cat lives to none.
By running, cat seems to know,
Aim **BANG!** Shot my SWOLLEN toe.

D.C.I.
(Delinquency Control Institute)
(Arizona State University 1958)

Took this special new course wondering why,
Juveniles cheat, rob, steal, murder or lie.
No experience working with children before,
Know little about them, want to learn more.

Coming to class each day with an open mind,
Answers to many questions, seeking to find.
Noted instructors prominent in their field,
Class not hardheaded & if wrong will yield.

The instructors have presented their facts.
Of some causes that leads to criminal acts.
They have also presented textbook theories,
Disputed by experienced students inquiries.

Serious discussions at split coffee breaks,
Some class members admitting past mistakes.
You wanted a second cup, but what the hell,
Get back to class because there's the bell.

Faculty assignments demand written reports,
Give the class numerous tests of all sorts.
Don't you know there is only a 24-hour day?
With job and school leaves no time to play.

Your lonesome family tries to remember you,
My boss requests case reports that are due.
Getting very grouchy and lose some friends;
When schools out, I'll have to make amends.

Be damned that dreaded day's actually here,
For 12 weeks I worked to improve my career.
Quest any information for my brain to cram,
For today's judgment day with a final exam.

Finally after the anxiety, tears and sweat,
Very pleased to learn the grade I will get.
Male and female our entire 46 member class,
Civilian and law-enforcement, all did pass.

Then go to the D.C.I. Banquet to celebrate,
Graduation of Nineteen Hundred Fifty-Eight.

Next day the pressure's off going to relax;
Boss is satisfied and I didn't get the axe.

Thinking back without a regret nor remorse,
In fact I'm glad I took this D.C.I. course.
Then it hits you and you give a happy sigh,
What do you know now I'm one of the alumni.

CONNECTED
(X-RATED)
(Narrator-Cowboy-Cowgirl)

Cowboy hatband, turquoise & silver sheen,
New denim smell, boots saddle soap clean.
Bola tie protruding western shirt collar,
Is cowboy payday! Time to whoop & holler!

Empty front pockets to pockets upon hips,
Non-obstruction while doing dancing dips.
Move belt buckle two belt loops to right,
No cold buckle, omit bare midriff plight.

Slim, trim cowgirl companion is his hope,
Barhop, exploiting essence of lava soap,
Toilet paper Bandaids dot razor cut face,
Bloodshot eyes prospecting style & grace.

Elbowing bar, add raw egg & salt my beer,
Hand on my shoulder, female voice I hear.
"Hi cowboy! I'm unattached take a chance?
Put your arms around me, hug me & dance."

Smiling we embrace and consort musically,
Join together by emotions and physically.
"Cowboy your pants pocket, what is hard?"
Um-oh-ah my pocket knife I highly regard.

"Cowboy! Pocket knife now grows in size!"

"Cowboy! Pocket knife strokes my thighs!"

"Cowboy! Pocket knife developing twitch!"

"Cowboy! Pocket knife stirs female itch!"

As dancing partners we are perfect blend,
Bar closing, leave with new found friend.
NO-TELL HOTEL, bedding as husband & wife,
With her whetstone, hone my pocket knife.

(Option)
With my whetstone, hone his pocketknife.

RETRIBUTION
(Narrator-Spirits-Mother Nature)

Indian Spirits, weary & defeated in fight,
Observe scars of wars named miners blight.
Great Mother of Earth, the spirits appeal,
The wounds to our environment please heal.

Mother Nature replies in voice of concern,
Edict! Mortals & Spirits mandate to learn,
What is raped from me can NOT be replaced,
Non-Indian curse, PAY FOR EARTH DISGRACED!

Revenge endowed to Arizona Indian Nations,
HOW? Gambling legalized upon Reservations.
180-degree reversal, ways and days of old,
Tribal Casinos now........................

2002

Sweet Leslie granddaughter of mine,
Now is high school graduation time.
12 years of education did complete,
Alone & as adult adjust to compete.

Grades & character beyond reproach,
By invitation BYU-East do encroach.
In life, en route being schoolmarm,
Pray! Dear Lord keep her from harm.

FISHING
(Embellished Fact)

My hobby's bass fishing and make no mistake,
Every opportunity I get I'm out on the lake;
Chiefly alone but occasionally take someone,
Fishing is relaxing, a heck of a lot of fun.

Launching my boat and making one final wish,
That I would catch and cull a limit of fish,
Straight as any arrow across a lake I drove,
Cast out and start fishing my favorite cove.

What's that long thing swimming in the lake?
Well I'll be darned it's a true rattlesnake!
Betting that the chance's definitely remote,
That any snake could ever get into the boat.

Hey wait a minute! Now it is heading my way!
Get out of here snake! I don't want to play!
What am I going to do? I am without any gun!
I am in the middle of a cove! Unable to run!

Had you been there would have laughed at me,
For I know it was a darn funny sight to see;
Grab one oar and hit that snake on the head,
Struck and struck until that snake was dead.

The oar was rotating from waterline to keel,
Imitating Mississippi riverboat paddlewheel;
No one on the lake either by sail or by gas,
Wasn't any kind of a problem for me to pass.

Return overworked oar to its original space;
Continue fishing in totally different place;
Even though I tried I didn't catch fish one,
For today's fishing I have had it! I'm done.

Loaded up my boat quietly and without delay,
If I come home without fish, what can I say?
This is not the kind of tale I like to tell;
Fish market, what kind of fish do they sell?

5-15-1966 ..... **5 SHOTS** ..... 4 HITS

KPD officer routed to a family fight,
Triggering sequence this fatal night.
Shot before turning off car ignition,
Is hospitalized in serious condition.

Graveyard-quiet and moonlight is nil,
In vacant field off of Stockton Hill.
Shooter retreats from house to field,
Freedom, choose to defend! Not yield.

MCSO deputies back KPD shooting call,
CAPTURE! Is recklessly sought by all.
An unanswered gunshot; anyone to die?
KPD Sergeant forfeits sight of 1 eye.

On point, all of other uniforms rear,
Mistaken for shooter is a great fear.
Where he is positioned I do not know,
Uncertain resolve is cautiously slow.

Use flashlight and create mortal sin,
Shooter would bereave my next of kin.
Reduce risk, yell FREEZE, HOLD SCENE!
REQUEST LIGHT AREA BY HEADLIGHT BEAM!

Now, wait a car being brought around,
Shadowy figure rises from the ground.
EXPLOSION! **FLASH** from his gun muzzle,
Presents final solution to my puzzle.

Shooter's **FLASH** highlight reflection,
Reply 2 shots in shooter's direction.
Deny his bullet. He has both of mine,
Shooter terminated in a fast recline.

Quest void! By silence of either gun,
Results? Everybody lost. Nothing won.
Car lights arrive, fears melted away,
Vivid 50 years, trauma burdens today.

5-15-1966, PEACE OFFICERS MEMORIAL DAY, **5-15-2016**
KPD Kingman Police Department officer Robert Wilkins and
Sergeant James McFerrin
MCSO Mohave County Sheriff's office-Deputy
Austin Cooper Backup and investigating officer

Deputy Phil Jordan (Future Sheriff #30) on Point

Law Enforcement responded to shooter's actions this night and I take no Pride for the end results. My sympathy for shooter's family and empathy for my brother officers and their families.

ARIZONAN

We are all gathered here this landmark day;
Tributes to "Good Ole Boy" who passed away.
We each knew him in a very different light;
List of his accomplishments? How to recite.

>Drank all the beer ever canned,
>Serviced all women in the land.
>Total singer, dancer and actor;
>Strong! Could out pull tractor.
>
>Perfect husband past six wives,
>A hero he saved multiple lives,
>Won every fight he ever fought,
>Teaching anything to be taught.
>
>Crossed any ocean as a swimmer,
>Climber of mountains is winner.
>Handsome, suave & was not vain,
>Captain submarine, ship, plane.
>
>Always bowling a perfect score,
>Conversationalist, not to bore.
>Sports first team All American,
>Each son greater than SUPERMAN.
>
>Miss America all his daughters,
>Trophy fish salt, fresh waters.
>As hunter, rewrote record book,
>Not any meal he could not cook.
>
>Best painter, sculptor, writer,
>A gastronomer, needless dieter.
>Broke no law, committed no sin,
>Never any bet he could not win.
>
>Could spout jokes for evermore,
>Singlehandedly winning any war.
>Pickup trucks king of the road,
>Was cowboy never to be throwed.
>
>In demand as political adviser;
>Intelligence? No one was wiser.

>   Going to Heaven withstood heat,
>   Income tax he refused to cheat.

Lets bring the eulogy to a screeching halt,
Whatever he fancied not entirely his fault.
Died of bowel obstruction, silencing a wit,
Good Ole Boy was and still is full of it.

LIGHTS OUT

San Carlos Apache Police two officers will do,
Maricopa, Mohave Sheriff's, each deputies two,
Brother officers pallbearers shouldering load,
First part of my journey upon the golden Road.

Popular or Western music 1940's, 50's or 60's,
Family, relatives, friends, minus dignitaries;
Smell the beautiful flowers arranged expertly,
Large middleclass casket color light mahogany.

Levi Jacket & pants, western shirt & bola tie,
Remember me thru my poetry, enjoy and goodbye.
WWII Combat Veteran and entitled burial rites,
2 Military bugles, taps, darken all my lights.

Dimple; fair skinned, blonde hair, mega nosed,
Clear blue eyes that saw too much, now closed.
Love my Church, thanks to those who did speak,
Thoughtful remembrances, life is now complete.

Veteran's Memorial Cemetery in plot of ground,
Located in Arizona, and could easily be found.
Now there is another challenge, new job to do,
Father in Heaven, hopeful to better serve you.

POETS POEM

```
*  Rhythmical, imaginative, uniquely concentrated;   *
*  Unlimited thoughts endowed to language related;   *
*  Multi words allocated in classic verse & rhyme;   *
*  Birth poems hopeful to withstand sands of time:   *
```

Poetic poem definition as many a person will know it,
Alas, absent of heart, soul & penmanship of the poet.
All dictionary words destined to remain alphabetical,
Till our concentration set them active & symmetrical.

Poets create for fun, some for fortune, few for fame,
And there are those that won't pen their proper name.
Fortunately poets carry banners of all walks of life,
Independently each march to a different drum or fife.

If you enjoy poetry and your emotions are stimulated,
This is the poem's intent that the poet contemplated.
Whether you like our poems or consider them disaster,
We would continue to write as a poet or as poetaster.

ABOUT THE AUTHOR

Walter Phillip Jordan Jr., better known to friends and loved ones as "Phil," created this book of heartfelt poems to offer hope to thousands of veterans struggling with post-traumatic stress disorder (PTSD).

After suffering with PTSD for many years, 91-year-old Phil felt passionate about giving back through this book. His poems range in topics from humorous to dark, touching on his struggles and triumphs in a way that is both personal and relatable. All of the poems were composed throughout his life, including during his time while serving in the U.S. military. Each piece that works together to create this book is a true expression of his experiences.

His extraordinary career extended across both the U.S. Navy and U.S. Army during WWII as well as in law enforcement. His masterful use of language creates a backdrop that readers will find fascinating. In one of his poems, he tells his family that his PTSD developed during his military and law enforcement service and "burdened him as a man, camouflage[d with] trauma."

At 15 years old, Phil ran away from home to join the military but was told that he was too young to enlist, so he began work in the shipyards as a coppersmith. After he turned 16 years old, he started work cranking planes for Chinese Army cadets during WWII. In 1944, just twenty days after turning 17, he enlisted in the U.S. Navy during the heat of battle in the South Pacific. He served on the D.E. #443 Kendall C Campbell and watched as the flag was raised at Iwo Jima. He was present at Tokyo Bay at the signing of the Japanese surrender.

Phil earned five Battle Stars by the time he left the Navy at 18 years old. Only a year later, he made the decision to re-enlist in the U.S. Army and spent three years as a paratrooper with the 82nd Airborne Division — where he was often tasked with bringing fallen brothers home.

In 1950 Phil returned home to Phoenix, Arizona and became a sheriff's deputy. In 1969, he graduated from the FBI's National Academy. He became a federal investigator at a Native American reservation and was later elected to be sheriff. After 22 years, he retired from law enforcement. During the 1980s, his latent PTSD surfaced. He found that writing poems and fishing helped him to "combat the ghosts of the past."

Phil passed away shortly after contracting Tactical 16 to publish his book. Per Phil's wishes, all proceeds from the book will be donated to the Wounded Warrior Project.

ABOUT THE PUBLISHER

TACTICAL 16

Tactical 16 Publishing is an unconventional publisher that understands the therapeutic value inherent in writing. We help veterans, first responders, and their families and friends to tell their stories using their words.

We are on a mission to capture the history of America's heroes: stories about sacrifices during chaos, humor amid tragedy, and victories learned from experiences not readily recreated — real stories from real people.

Tactical 16 has published books in leadership, business, fiction, and children's genres. We produce all types of works, from self-help to memoirs that preserve unique stories not yet told.

You don't have to be a polished author to join our ranks. If you can write with passion and be unapologetic, we want to talk. Go to Tactical16.com to contact us and to learn more.

www.ingramcontent.com/pod-product-compliance
Lightning Source LLC
Chambersburg PA
CBHW042112120526
44592CB00042B/2725